Frederick Stearns Hollis

A Further Investigation of the Chlorides of

Paranitroorthosulphobenzoic Acid

Frederick Stearns Hollis

A Further Investigation of the Chlorides of Paranitroorthosulphobenzoic Acid

ISBN/EAN: 9783337339678

Printed in Europe, USA, Canada, Australia, Japan

Cover: Foto ©berggeist007 / pixelio.de

More available books at **www.hansebooks.com**

Contents

Acknowledgment

I. Introduction
II. Preparation of Material
III. The Action of Phosphorous Pentachloride on the Acid Potassium Salt of Paranitro-orthosulphobenzoic Acid.
   a. Preparation of the Unsymmetrical Chloride
   b. Preparation of the Symmetrical Chloride.
IV. The Action of Benzene and Aluminium Chloride on the Chlorides of Paranitroortho-sulphobenzoic Acid.
   a. The Action on the Unsymmetrical Chloride
   b. The Action on the Symmetrical Chloride.
V. The Action of Hydrochloric Acid on Paranitroorthobenzoylbenzenesulphone Chloride
   a. The Action of Dilute Hydrochloric Acid.
   b. The Action of Concentrated Hydrochloric Acid.

VI. The Action of Dilute Sulphuric Acid on Paranitroorthobenzoylbenzenesulphone Chloride

VII. The Action of Water on Paranitroorthobenzoylbenzenesulphone Chloride.

VIII. The Action of Absolute Alcohol on Paranitroorthobenzoylbenzenesulphone Chloride.

IX. Comparison of the Barium Salts of Paranitroorthobenzoylbenzenesulphonic Acid.

X. Preparation of Other Salts from the Barium Salt.
   a. The Sodium Salt.
   b. The Potassium Salt.
   c. The Magnesium Salt.
   d. The Calcium Salt.
   e. The Lead Salt.

XI. The Action of Phosphorus Pentachloride on the Sodium Salt of Paranitroorthobenzoylbenzenesulphonic Acid.

XII. The Action of Concentrated Ammonia

Paranitroorthotoluenebenzenesulphonic chloride
a. The Action of Ammonia on the chloride in an Open Vessel.
b. The Action of Ammonia on the chloride in a Sealed Tube.
- The Action of Concentrated Ammonia on the Lactim of Paranitroorthobenzoylbenzene-sulphonic acid.
- The Action of Dilute Hydrochloric Acid on the Lactim of Paranitroorthobenzoyl-benzenesulphonic Acid.

Conclusion.

Biographical.

Acknowledgment

This investigation was suggested by Professor Ira Remsen, and the work carried on under his supervision. To him the author wishes to express his gratitude for guidance in this work and for instruction received in the lecture room. He also wishes to express his appreciation of the instruction received from Professors Morse and Clarke and Dr. Mathews.

I Introduction

The present investigation may be divided into two parts. The first consists of further work on the method of preparation and separation of the chlorides of paranitroorthosulphobenzoic acid. This work was confined largely to the unsymmetrical or low-melting chloride as this is the one used mainly in the second part of the investigation. The preparation of the unsymmetrical dichloride, first obtained by Gray, was found to be a matter of considerable difficulty and uncertainty, unless crystallization could be conducted out of doors at a very low temperature. As the result of a series of experiments undertaken to determine the best conditions for the preparation of the unsymmetrical chloride, a method has been adopted by which the unsymmetrical chloride may be prepared in any desired quantity in the laboratory.

The second part of the investigation consists of a study of the action of benzene and aluminium chloride on the chlorides under varying conditions, and the preparation of a series of derivatives of the product formed. Saunders' and McKee,[2] working with the chlorides of orthosulphobenzoic acid,[1] found that the action of benzene and aluminium chloride gives the same product with both chlorides.

It was thought that, on account of the greater stability of the paranitroorthosulphobenzoic acid, the action of benzene and aluminium chloride might lead to the formation of two series of derivatives, one derived from the symmetrical chloride, and the other from the unsymmetrical chloride, in which the resulting compound would retain the unsymmetrical structure of the chloride. This proved not to be the case, as the product of the action of benzene and

Aluminum chloride on both of the chlorides was found to be paranitroorthobenzoylbenzenesulphone chloride.

It was found that paranitroorthosulphobenzoic acid does not yield a sulphone corresponding to the one obtained from orthosulphobenzoic acid by Saunders', although the reaction was conducted under the conditions used by him, as well as under a variety of different conditions.

A series of derivatives of paranitrobenzoylbenzenesulphone chloride were prepared.

II Preparation of Material

The acid potassium salt of paranitroortho-sulphobenzoic acid was prepared from paranitrotoluene according to the method described by Kastle[1], and used later by Gray[2]. This consists in introducing the sulphonic acid group into paranitrotoluene and passing through the calcium salt to the potassium salt of paranitrotoluene-orthosulphonic acid. The potassium salt is oxidized by means of potassium permanganate to the potassium salt of paranitroortho-sulphobenzoic acid, and this is converted into the acid potassium salt by the action of hydrochloric acid.

1000 grams of paranitrotoluene gave 1439 grams of the neutral potassium salt of paranitrotoluene-orthosulphonic acid. 1000 grams of the potassium salt of paranitrotolueneorthosulphonic acid gave 800 grams of the acid potassium salt of paranitroorthosulphobenzoic acid.

[1] A.C.J. XI-177. [2] Inaug. D. Johns Hopkins University 1.

III. The Action of Phosphorus Pentachloride on the Acid Potassium Salt of Paranitroorthosulphobenzoic Acid.

The action of phosphorus pentachloride on the anhydrous acid potassium salt of paranitroorthosulphobenzoic acid gives rise to the formation of an unsymmetrical and a symmetrical dichloride, as determined by Gray, according to the following equations:

$$C_6H_3\begin{cases}COOH\\SO_2OK\\NO_2\end{cases} + 2\,PCl_5 = C_6H_3\begin{cases}CCl_2\\SO_2\\NO_2\end{cases}\!\!>O + 2\,POCl_3 + HCl + KCl$$

$$C_6H_3\begin{cases}COOH\\SO_2OK\\NO_2\end{cases} + 2\,PCl_5 = C_6H_3\begin{cases}COCl\\SO_2Cl\\NO_2\end{cases} + 2\,POCl_3 + HCl + KCl$$

Gray found that the relative amount of each chloride formed depends on the temperature and length of time which the phosphorus pentachloride is allowed to act on the acid potassium salt. The largest amount of the unsymmetrical chloride amounting to 80 %

per cent of the total, was formed by heating a mixture of two molecules of phosphorus pentachloride and one molecule of the anhydrous acid potassium salt to 150°C, in a distilling-bulb immersed in a sulphuric-acid bath, for four or five hours.

Under these conditions, the amount of the symmetrical chloride formed amounted to 0 or 20 per cent of the total, but this is increased to 30 per cent by heating in an open dish on a water-bath. The symmetrical chloride was separated by using chloroform as a solvent.

As considerable difficulty was experienced, mainly in the preparation of the unsymmetrical chloride, according to the directions given by Gray, a series of experiments was made under different conditions with amounts of the acid potassium salt varying from twenty to sixty grams in order to determine the con-

ditions most favorable for the formation of the unsymmetrical chloride. The results of these experiments are embodied in the following method.

a. The Preparation of the Unsymmetrical Chloride.

A mixture of one molecule of the acid potassium salt, previously heated to 150°C for four hours, and two and one half molecules of phosphorus pentachloride is carefully ground together in a mortar, and introduced into a distilling-bulb, having a capacity of about six times that of the volume of the mixture. The outlet-tube of the bulb is closed, and a cork, through which runs a glass tube about three feet long, the lower end reaching nearly to the surface of the mixture, inserted in the neck.

The distilling-bulb, thus closed, is immersed in a sulphuric-acid bath.

heated to 150°C, and this temperature maintained for five hours.

No reaction takes place on mixing the acid potassium salt and the phosphorus pentachloride, but, upon immersing the bulb containing the mixture in the heated bath, vigorous action begins immediately, and the resulting phosphorus oxychloride is conducted back upon the product by means of the condensing-tube. The temperature of 150°C cannot be safely exceeded, as decomposition of the chlorides begins at 160°C.

At the end of five hours the tube is removed from the neck of the flask, the perforated stopper replaced by a solid one, the outlet-tube opened, and the phosphorus oxychloride distilled off.

The resulting chloride, which is in the form of a thick, yellow, oily liquid,

is poured into a large bottle, nearly filled with cold water, and shaken vigorously so as to break it into small globules. This washing is continued until the chloride hardens to a solid cake, which commonly takes place after washing with five or six successive portions of cold water. The solidified chloride is broken up and dried by pressing between filter-paper and, in this form, it may be exposed to the air without undergoing much, if any, decomposition.

By using two and a half molecules of phosphorus pentachloride, all of the acid potassium salt is converted into the form of the chloride, while, if but two molecules are used, varying amounts of the acid potassium salt, in some cases as much as 25 per cent, are unacted upon and may

be recovered from the wash-water.

The product obtained by this method consists, with the exception of slight impurities of only the unsymmetrical chloride, thus making the crystallization from chloroform unnecessary. 100 grams of anhydrous acid potassium salt give, on an average, 99 grams of the crude unsymmetrical chloride.

Previous work showed that the crystallization of the unsymmetrical chloride from ligroin (50-80°) was a difficult matter, unless it could be done out of doors during very cold weather.

The same was found to be the case with the unsymmetrical chloride of orthosulphobenzoic acid by Saunders.

A solution of the crude chloride in purified ligroin (50-80°), on standing out of doors at a temperature of 6°F., crystallized in clusters of crystals some of which measured three centi-

meters in length. An attempt was made to obtain the chloride in the form of crystals by cooling the ligroin solution to 0°C. in a refrigerating box such as was used by The Hu.' The chloride invariably separated as an oil, and no advantage was derived by drawing a current of cold dry air through the flask containing the ligroin solution. Some of the oily chloride which separated from the ligroin solution, and from which the ligroin had been decanted, formed an opaque semi-solid mass on placing it in a freezing-mixture, but no crystals were deposited.

The first indication of crystallization in the laboratory was obtained on placing some of the oil, which had several times been dissolved in fresh portions of hot ligroin (50-80°) and allowed to separate out on cooling, in a freezing-mixture and stirring with a rod.

Inaug. Diss. Johns Hopkins Univ. 1895.

This suggested the possibility that some material which by its presence retards crystallization, had been dissolved out of the mass by the several portions of ligroin in which it had been dissolved. This was shown to be the case by several tests and gave rise to the following method of purification and crystallization, in which the further change is made of using ligroin of boiling point 90-125° as the solvent. This ligroin is to be preferred to that having a lower boiling-point, as the chloride is apparently much more soluble in it and crystallizes from the solution at the temperature of the laboratory.

The ligroin is purified by shaking in a separating funnel with concentrated sulphuric acid until it imparts no color to a fresh portion of acid, after which it is treated with caustic soda solution to neutralize the acid, and washed free from alkali by

The crude chloride, which is always somewhat dark colored and gummy, is placed in an Erlenmeyer flask with purified ligroin (50-80°) and washed by stirring it down with a rod until it yields a granular powder having but little color.

On boiling the chloride thus purified with ligroin (90-125°), it dissolves, with the exception of slight remaining impurities, and on cooling the solution becomes cloudy and the excess of chloride separates out as a light colored oil. It is best to decant the solution from the separated oil after separation has mainly ceased at the temperature at which crystallization is to proceed, but before the solution has become clear. The chloride is obtained from this solution at the temperature of the laboratory in clusters of needles having the form of long monoclinic prisms as observed

by Gray[1]. The rate of crystallization is increased by keeping the solution at a lower temperature, but no especial advantage is derived unless the temperature is very low, when larger crystals are obtained.

The crystals of the chloride thus obtained have a constant melting-point of 57°C (uncor.)

The purified chloride which separates from the ligroin as an oil generally crystallizes on standing, but it is better to redissolve it in a fresh portion of ligroin, by which it is further purified, and proceed according to the directions given above.

The principal impurity which causes difficulty in the crystallization of the unsymmetrical chloride seems to be that which is removed by the preliminary treatment with ligroin (50-80°).

On evaporating the ligroin used in washing the chloride, this impurity remains as a

[1] Inaug. Diss. Johns Hopkins University 1895

dark colored slightly viscous liquid having a strong acid reaction. It showed no tendency to crystallize after standing in the laboratory for five months. A small amount of another impurity was obtained as a flocculent material on dissolving the crude chloride in chloroform. It melts after purification at 100-105°C, and is probably the anhydride.

7. The Preparation of the symmetrical Chloride.

One portion of the symmetrical chloride was made in order to test the action of benzene and aluminium chloride, but no comparative tests were made, as in the case of the unsymmetrical chloride. The method used for its preparation differed somewhat from that of Gray.

The conditions chosen for its preparation were, as far as possible, the opposite of those found to be most favorable for the formation of the unsymmetrical chloride.

The anhydrous acid potassium salt and phosphorus pentachloride, in the proportion of one molecule of the former to two molecules of the latter, are mixed by grinding them together, and the action commenced by placing the vessel containing the mixture in a sulphuric-acid bath, previously heated to 150°C.

The vessel is removed from the bath as soon as the action commences, and allowed to stand until the action is complete, which requires about ten minutes.

The mass is washed, as in the case of the unsymmetrical chloride, by shaking in a bottle with five or six portions of cold water.

The chloride thus washed exists in the form of a light colored, thick gum. It is dissolved in anhydrous chloroform, the solution dried by means of calcium chloride and allowed to stand at the temperature of the laboratory. Crystals of the symmetrical chloride are formed only after the chloroform has nearly all evaporated, and crystallization proceeds very slowly.

The chloride was obtained in the form of small monoclinic crystals, having a constant melting-point of 94°C. (uncor.)

The yield indicated that, under the conditions used, the action of the phosphorus pentachloride is incomplete, but between 35 and 40 per cent. of the product obtained consists of the symmetrical chloride.

IV. The Action of Benzene and Aluminium Chloride on the Chlorides of Paranitroortho-sulphobenzoic Acid.

a. The Action on the Unsymmetrical Chloride.

When aluminium chloride is added to a solution of the unsymmetrical chloride in benzene, slight action begins immediately, as is shown by the darkening of the color of the solution and a slight evolution of hydrochloric acid gas.

In the case of small quantities of the chloride, heated with an excess of aluminium chloride, the reaction is complete in from fifteen to twenty minutes. A series of experiments was made in order to determine the best conditions for conducting the reaction and for separating and purifying the product.

Under some conditions the product contains a considerable amount of a solid.

material which separates as an impurity on crystallization. As a result of these experiments, the following method was adopted:

Twenty grams of the unsymmetrical chloride is dissolved in one hundred cubic centimeters of benzene in a flask provided with a return-condenser, and about ten grams of aluminium chloride in small pieces added.

This is heated with a small flame for from one to two minutes at the end of which time vigorous action commences and continues without further heating for ten minutes.

After the action is over, the flask is heated repeatedly so as to maintain an even evolution of hydrochloric acid gas for about eight minutes, at the end of which the reaction is complete. The resulting product is poured into seven hundred and fifty cubic centimeters of water in a liter separating-

funnel, seventy five cubic centimeters of hydrochloric acid (sp. gr. 1.12) are added and the mixture well shaken.

After the benzene layer has risen, the water is drawn off, and the small amount of the product suspended in it separated by filtration. The greater part of the product is in the form of a pinkish-white powder, which remains in suspension in the benzene, from which it is separated by filtration, dried, and purified with the portion obtained from the water. The product is purified by dissolving in benzene and adding rather more than an equal volume of anhydrous ether, which causes a more rapid crystallization.

By this method it is obtained in clusters of small crystals, having a rhombohedral habit or as a granular powder if crystallization takes place rapidly

Some of the larger crystals measured four or five millimeters on an edge. The larger crystals have a purple or greenish color, and the granular form is generally slightly green. Both forms yield a white powder.

The pure crystalline product has a constant melting-point of 177°C. (uncor.)

The portion of the product which remains in solution in the benzene is mixed with a small amount of the purple impurity before described, but it is obtained in a crystalline form of fair purity by drying the benzene solution, evaporating it to about one third of its volume, and adding rather more than an equal volume of absolute ether. The dark impurity may be largely removed by washing with absolute alcohol in which it dissolves readily.

An attempt was made to prevent the dark-

ening of the benzene solution during the evaporation by conducting the evaporation in a current of sulphur dioxide, as it was believed that the darkening was due, in part at least, to the action of the air.

The product was not materially improved and a certain amount of free acid was always found to be present after such treatment. The rapid evaporation of the dried benzene solution and the addition of an equal volume of absolute ether to ensure rapid crystallization is greatly to be preferred.

Twenty grams of the unsymmetrical chloride gave sixteen grams of the product as first obtained. This was slightly decreased by recrystallization.

The product has a characteristic disagreeable odor.

7. The Action on the Symmetrical Chloride.

The reaction was conducted exactly as in the case of the unsymmetrical chloride, using the same relative quantities and method of treatment. The method of separation was also the same.

The product gave, on purification, crystals of the same form, size, and color as those obtained from the unsymmetrical chloride.

The melting-point of the purest crystals is also the same, 77°C. uncor.

Two grams of the symmetrical chloride gave one and a half grams of the product.

Analyses of the Product of the Action of Benzene and Aluminium Chloride on the Unsymmetrical Chloride.

0.2051 gram gave	0.3632 gram	CO_2
0.2992	0.5249	
0.1994	0.0535	H_2O
0.2992	0.0819	
0.2508	9.03 cc	n
0.3496	12.88 "	"
0.2601	0.1957 gram	$BaSO_4$
0.2058	0.1584	
0.2602	0.1818	
0.2492	0.1878	
0.2424	0.1811	
0.2434	0.1072	$AgCl$
0.2058	0.0925	
0.2602	.1	

Calculated for $C_{13}H_8O_5N_2SCl$		Found					
		1	2	3	4	5
C	47.92	48.29	47.84	—	—	—
H	2.45	2.97	3.04	—	—	—
N	4.30	4.32	4.42	—	—	—
S	9.83	10.33	10.56	9.96	10.34	10.26
Cl	10.90	10.90	11.11	10.03	—	—

The two following structural formulae are possible for a substance derived from the unsymmetrical chloride, and having the composition indicated by the analyses.

$$C_6H_3 \cdot NO_2 \left< \genfrac{}{}{0pt}{}{C_6H_5}{SO_2} \right> O \qquad C_6H_3 \cdot NO_2 \left< \genfrac{}{}{0pt}{}{CO \cdot C_6H_5}{SO_2 \cdot Cl} \right.$$

The fact that the product is apparently not acted upon by alcoholic potash, together with its high melting-point and its properties generally favor the belief that it has the structure represented by the first.

The formation of the same product by the action of benzene and aluminium chloride

on the symmetrical chloride seems to indicate that the product derived from each chloride is paranitroorthobenzoylbenzene sulphone chloride. This new agrees with the results obtained by the action of benzene and aluminium chloride on orthosulphobenzoic acid by Saunders and Mc Kee².

It is clear from the analyses that but one of the chlorine atoms of the chloride is replaced by this reaction. All attempts to prepare the di-phenyl derivative of paranitroorthobenzoyl-benzene sulphone were unsuccessful, although the conditions were varied widely, both as to temperature and the length of time which the heating was continued.

The conditions already described give the best yield and also the purest product.

By allowing the mixture to stand for a day with occasional heating nearly to the boiling,

¹ A. C. J. XVII 355   ² Inaug. Diss. Johns Hopkins University 1895.

point of the benzene, similar good results are obtained.

By heating to the boiling-point of benzene, for three hours, using a return-condenser, the yield is decreased, and a large amount of a black, tarry matter is obtained which is almost insoluble in benzene. This dissolves readily in absolute alcohol, from which solution it is precipitated as a dark red-colored powder on adding water and, after being thrown out of solution in this way, it becomes less soluble in absolute alcohol. It swells up on heating and gives an odor like that obtained on burning sulphonic acids. On burning off the organic portion a considerable amount of alumina remains. Hydrochloric acid dissolves the alumina, leaving the organic portion in the form of a black, tarry mass

II. The Action of Hydrochloric Acid on Paranitroorthobenzoylbenzenesulphone Chloride.

a. The Action of Dilute Hydrochloric Acid (Exp. 12)

The action of dilute hydrochloric acid was determined by boiling the sulphone chloride in a flask, provided with a return-condenser, with an excess of the acid until it was all dissolved, which usually requires about six hours. The solution is then filtered and evaporated on a water-bath, and the heating continued until no odor of hydrochloric acid remains. The resulting acid is obtained in the form of a dark solid substance which dissolves readily in water and takes up water on standing in the air.

Preparation of the Barium Salt. — The barium salt was prepared by adding barium carbonate to a solution of the acid in water,

filtering off the excess of barium carbonate, and evaporating the solution under a bell-jar by means of a current of dry air. The solution cannot be safely concentrated by boiling, as it causes a decomposition of the salt.

The barium salt was obtained in the form of small light-colored crystals, arranged in tufts. In the following analyses of salts the base, as well as the sulphur, is calculated on the basis of the anhydrous salt.

0.2039 gram lost 0.0119 gram at 180°C. and gave 0.0586 gram $BaSO_4$
0.2892 "   "   0.0179 "   "   "   "   "   0.0810 "   "

	Calculated for $(C_{13}H_{10}O_6 \cdot 2S)_2 Ba + 3H_2O$	Found 1	2
H_2O	6.72	5.84	6.18
Ba	18.29	18.12	17.6?

The barium salt of another portion of acid

prepared in the same way was obtained in the form of short, thick monoclinic prisms, which seemed to be made up of a series of plates.

0.2010 gram lost 0.0260 gram at 210°C. and gave 0.0557 gram Ba.
0.2021 "   " 0.0256 "   "   "   "   " 0.0562 "   "
0.2263 " gave 0.1262 gram BaSO$_4$
0.2109 "   " 0.1201 "   "

Calculated for $(C_{13}H_8O_6 \cdot S \cdot Cl)_2 Ba + 6H_2O$		Found	
		1	2
H_2O	12.60	12.93	12.66
Ba	18.29	18.14	18.70
S	8.54	8.76	7.

This salt became opaque on standing in a specimen tube for one month, due to loss of water of crystallization.

0.2133 gram of the opaque salt lost 0.0139 gram at 210°C.

Calculated for $(C_{13}H_8O_6 \cdot S)_2 Ba + 3H_2O$		Found
H_2O	6.72	6.51

The results of analysis indicate that para-nitroorthobenzoylbenzenesulphone chloride is converted into paranitroorthobenzoylbenzene-sulphonic acid by boiling with dilute hydrochloric acid, according to the following equation.

$$C_6H_3NO_2\genfrac{<}{}{0pt}{}{COC_6H_5}{SO_2Cl} + H_2O = C_6H_3NO_2\genfrac{<}{}{0pt}{}{COC_6H_5}{SO_2OH} + HCl$$

5. The Action of Concentrated Hydrochloric Acid (sp. gr. 1.17)

The action of concentrated hydrochloric acid was determined by heating the sulphone chloride with a large excess of acid in a sealed tube.

The tube was first heated for six hours in a water-bath but, as no action seemed to take place, it was transferred to a Carius furnace and heated for six hours at a

temperature of 175°C. The substance dissolved, with the exception of a few dark flakes, and the acid was colored brown. The flakes were removed by filtration, the acid solution evaporated to dryness on a water-bath, and the heating continued until the resulting acid had no odor of hydrochloric acid.

The barium salt was prepared as in the case of the acid derived from the sulphone chloride by the action of dilute hydrochloric acid. It crystallized in the form of light-colored, fine needles, which were arranged in loose tufts or clusters. A few darker crystals in the form of larger monoclinic crystals with ~~~~~~~ were obtained from the mother-liquor.

0.2103 gram of the needles lost 0.0164 gram at 210°C. and gave 0.609 gram $Ba SO_4$

0.2083 " " " " 0.0158 " " 0.0595 "

Calculated for $(C_{13}H_9O_6N S)_2 Ba + 3½ H_2O$ Found
                                                    1   2
H_2O 7.75   7.69   7.59
Ba 18.29   18.21   18.20

0.1053 gram of the larger, dark crystals lost 0.0152 gram at 210°C. and gave 0.0279 gram $BaSO_4$

Calculated for $(C_{13}H_9O_6N S)_2 Ba + 7 H_2O$ Found
H_2O 14.40   14.43
Ba 18.29   18.20

VI. The Action of Dilute Sulphuric Acid on Paranitroorthobenzoylbenzenesulphone Chloride.

The action of sulphuric acid on the sulphone chloride was determined by heating in a flask with a return-condenser until it dissolved. The resulting product was an acid, which was converted into the barium salt by adding an excess of barium carbonate as in the previous experiments.

The barium salt was obtained in the form of short needles arranged in clusters.

0.1943 gram lost 0.0119 gram at 150°C. and gave 0.0570 gram Ba SO$_4$

Calculated for $(C_{13}H_8O_6 N S)_2 Ba + 3H_2O$		Found
H$_2$O	6.72	6.1.
Ba		17.4

VII. The Action of Water on Paranitro benzoylbenzenesulphone Chloride.

The action was determined by boiling the sulphone chloride in a flask with a return-condenser until it dissolved. It was necessary to boil somewhat longer to dissolve the substance than in the experiments in which acids were used. The resulting product was an acid which, by treating in the usual way with barium carbonate, gave a barium salt which crystallized in well formed monoclinic crystals.

0.1610 gram lost 0.0207 gram at 210°C and gave 0.0443 gram $BaSO_4$.

Calculated for $(C_{13}H_8O_6NS)_2Ba + 6H_2O$		Found
H_2O	12.60	12.85
Ba	18.29	18.56

VIII. The Action of Absolute Alcohol on Para-nitroorthobenzoylbenz. sulphone Chloride

The action of absolute alcohol on the sulphone chloride was determined by boiling in a flask with a return-condenser until it dissolved. It dissolved rather more rapidly than in the experiments in which acid was used, and the boiling was continued for a short time after all the material was dissolved. After a part of the alcohol was evaporated, a few drops of the solution showed indications of crystallization on evaporating rapidly on a watch-glass but, on further evaporation, the solution darkened and the resulting product was an acid as in the previous experiments. The barium salt was prepared as in the previous experiments. The crystals first obtained were in the form of small light-

colored needles arranged in clusters but later well formed monoclinic crystals were obtained from the same solution.

0.1957 gram of the needle shaped crystals lost 0.0153 gram at 210°C. and gave 0.0559 gram of BaSO₄

Calculated for $(C_{13}H_8O_6NS)_2Ba + 3\frac{1}{2}H_2O$ Found

H₂O	7.75	7.81
Ba	18.29	18.21

0.2007 gram of the larger crystals lost 0.0285 gram at 210°C. and gave 0.0535 gram of BaSO₄

Calculated for $(C_{13}H_8O_6NS)_2Ba + 7H_2O$ Found

H₂O	14.40	14.26
Ba	18.29	18.27

The above analyses indicate that the action of dilute or concentrated hydrochloric acid, sulphuric acid, water and alcohol on para-nitroorthobenzoylbenzenesulphone chloride converts it into paranitroorthobenzoylbenzenesulphonic acid.

IX. Comparison of the Barium salts.

The analyses of the barium salt of the acids, derived from the action of various substances on the sulphone chloride, show that the acid is in every case the same and that the salts contain the same amount of barium when calculated upon the basis of the anhydrous salt. The amount of water of crystallization varies widely in the different salts, depending on the conditions under which crystallization takes place.

The needles are obtained from the more concentrated solutions, and crystals of this form are first obtained from a solution which has been evaporated by heating before placing it under a bell-jar. All crystals having this form contain three or three and a half molecules of water of crystallization.

The larger monoclinic crystals which form in the same solution after the formation of needles ceases, or when a cold solution is evaporated to the point of crystallization under a bell-jar, are characterized by six molecules of water of crystallization. On exposure to the air or even in a stoppered tube these lose water of crystallization and become opaque.

The only analysis made of a crystal that had changed in this way shows that it contains three molecules, while it crystallized with six.

Those crystals which contain seven molecules of water of crystallization are obtained on slow crystallization, on standing in the air, from a dilute solution or from a mother-liquor from which crystals containing a less amount of water of crystallization have been deposited. Although all the barium salts

containing different amounts of water of crystallization appear to crystallize in the monoclinic system, they show clearly a variation in form. Owing to lack of time, no comparative study could be made of the relation existing between the amount of water of crystallization and the crystallographic constants.

The barium salt of paranitroorthobenzoyl-benzenesulphonic acid is characterized by an intense bitter taste.

1. Preparation of Other Salts from the Barium
Salt of Paranitroorthoxy-azobenzenesulphonic
Acid.

These were prepared from an aqueous solution of the barium salt by precipitating the barium exactly by means of sulphuric acid, and neutralizing the free acid exactly with the carbonate of the base.

The solutions were evaporated to crystallization under a bell-jar by means of a current of dry air.

In the analyses the amount of the base is calculated on the basis of the anhydrous salt.

a. The Sodium Salt.

The sodium salt was obtained in the form of fine white crystals composed apparently of monoclinic prisms.

They appear to undergo no change on exposure

Li th...

0.1??? gra.   0.0.05 gram at 210°C. and gave 0.0384 gram Na_2SO_4
0.2022   "   0.0111   "   "   "   "   0.0402   "   "

Calculated for $(C_{13}H_8O_6 n S) Na + 1H_2O$ Found
                                              1   2
H_2O 5.19   5.53   5.49
Na   6.99   6.93   6.82

7. The Potassium Salt.

The Potassium salt was obtained in the form of fine white needles which were too small to indicate the form of crystallization. They became opaque on exposure to the air.

0.2118 gram lost 0.0014 gram at 210°C, and gave 0.0546 gram K_2SO_4
0.204   "   "   0.0013 "   "   "   " 0.0518 "   "

Calculated for $(C_{13}H_8O_6 n S) K$ Found
                                         1   2
K   11.33   11.6?   1?

c. The Magnesium Salt.

The magnesium salt was obtained in the form of tabular monoclinic crystals having a marked pearly lustre. Some of the crystals measured nearly a centimeter in length. They appear to undergo no change on exposure to the air.

0.1940 gram lost 0.0408 gram at 2 0°C. and gave 0.0294 gram Mg SO$_4$
0.1997 "   " 0.0425   "   " 0.0303

Calculated for (C$_{12}$H$_8$O$_6$)$_2$ 5/2 Mg + 9.2 H$_2$O,   Found
                                                            1   2
H$_2$O   21.17   21.03   21.28
Mg   3.83   3.88   3.90

d. The Calcium Salt.

The calcium salt was obtained in the form of thin pearly plates having no regular bounding planes. They become opaque on exposure to the air and crumble to a white powder.

0.1373 gram lost 0.0096 gram at 210°C and gave 0.0280 gram $CaSO_4$

0.1199 "   "   0.0082 "   "   "   "   0.0242 "   "

Calculated for $(C_{13}H_8O_6 \cdot 2S)_2 Ca + 3H_2O$ Found

	Calculated	1	2
H_2O	7.65	6.99	6.88
Ca	6.13	6.44	6.39

The Lead Salt.

The lead salt was obtained in clusters of small tabular monoclinic crystals, which become opaque very slowly on exposure to the air.

0.2121 gram lost 0.0219 gram at 210°C, and gave 0.0711 gram $PbSO_4$

0.2039 "   "   0.0213 "   "   "   "   " 0.0699 "   "

0.1543 "   "   0.0166 "   "   "   "   " 0.0509 "   "

Calculated for $(C_{13}H_8O_6 \cdot 2S)_2 Pb + 5\frac{1}{2}H_2O$ Found

	Calculated	1	2	3
H_2O	10.78	10.32	10.44	10.75
Pb	25.25	25.43	25.95	25.23

The copper salt ... ... went decomposition on evaporation

XI. The Action of Phosphorus Pentachloride on the Sodium Salt of Paranitro orthobromo benzenesulphonic Acid.

The sodium salt and phosphorus pentachloride in the proportion of one molecule to one and a half were mixed by grinding together in an evaporating dish. There were no evidences of action, even upon adding a considerable quantity of phosphorus oxychloride, but on heating there was slight action. The heating was continued for about ten minutes, and the pasty mass was then treated with a considerable volume of cold water. Most of the material dissolved, but part hardened to a solid mass. After carefully washing with water, this material was washed with absolute alcohol, dissolved in benzene, and crystallized out by adding an equal volume of anhydrous ether. The product separated

out as clusters of small, light colored crystals which melted at 174-6°C (uncor.) and as a scale around the sides of the beaker, which melted at 160-170°C (uncor.) It was entirely free from the dark purple material obtained as an impurity in the preparation of para nitroorthobenzoylbenzene sulphone chloride by the action of benzene and aluminium chloride. A considerable portion of the material was insoluble in benzene and melted at 270-5°C (uncor.)

The method of formation of this material, together with its melting point, its solubility in benzene, from which it crystallizes readily upon the addition of absolute ether, indicate that it is paranitroorthobenzoylbenzene sulphone chloride. The method of formation from the sodium salt is indicated in the following equation:

$$C_6H_3NO_2\genfrac{}{}{0pt}{}{COC_6H_5}{SO_2ONa} + PCl_5 = C_6H_3NO_2\genfrac{}{}{0pt}{}{COC_6H_5}{SO_2Cl} + POCl_3 + NaCl$$

The material melting at 174-6° was boiled in a flask, provided with a return-condenser, with an excess of dilute hydrochloric acid until it was completely dissolved. This required seven hours. The solution was filtered and evaporated to dryness on a water-bath, and the heating continued until all hydrochloric acid was driven off. The product was a dark solid, similar to that obtained by the action of acid on paranitroorthobenzoyl-benzenesulphone chloride. An excess of barium carbonate was added to an aqueous solution of the product. The excess of carbonate filtered off and the solution, which had the characteristic bitter taste of the barium salt of paranitroortho-benzoylbenzenesulphonic acid, evaporated under a bell-jar. On evaporation the solution yielded a small amount of a crystalline product containing barium, which shows that the product is the barium salt of an acid.

The formation of para nitrothenzoylbenzene sulphonic acid by the action of hydrochloric acid on the product of the action of phosphorus pentachloride on the sodium salt of paranitro-orthobenzoylbenzene sulphonic acid, and its conversion into the barium salt, confirms the view already expressed that the action of phosphorus pentachloride on the sodium salt gives the sulphone chloride.

XII. The Action of Concentrated Ammonia on Paranitroorthobenzoylbenzenesulphone chloride.

a. The Action of Concentrated Ammonia on the Chloride in an Open Vessel.

Some of the sulphone chloride boiled with concentrated ammonia in a test-tube was apparently but little changed, but on filtering off the unchanged chloride and evaporating the ammonia, a yellow amorphous powder was obtained which melted at 234° (uncor) after washing with water to remove any ammonium chloride. The action of ammonia on the chloride was next tried by boiling in a flask, provided with a return-condenser to which fresh portions of ammonia were added as the strength decreased by evaporation. After boiling in this way for about three hours, the ammonia became yellow, although the

material in a mig undissolved was not changed in appearance. It was found, however, that the insoluble material melted above 334°C. uncor.¹ and that it was unchanged by crystallization from benzene and alcohol.

A small amount of material was also obtained which melted above 375°C uncor.¹ This indicated that the chloride had undergone a change on heating with ammonia.

3. The Action of Concentrated Ammonia on the Chloride in a Sealed Tube.

In the experiments in which the chloride was heated with concentrated ammonia in an open vessel, the strength of the ammonia decreased and, even though continually replaced, the experiment was complicated by the action of water. In order to avoid this the chloride was heated with concentrated am-

monia in a sealed tube. Upon heating as long as it had been found necessary to heat in an open vessel in order to effect the transformation, the product was mixed with a considerable amount of a red-colored material which melted above 275°C (uncor.)

As the result of several experiments it was found that, by heating the chloride in a sealed tube for two or two and a half hours in a water-bath, it was mainly converted into a clear, granular product which melted at 234°C (uncor.), as in the previous experiments.

The small amount of the dark high-melting product formed exists as a thin coating, and can easily be removed mechanically or dissolved in alcohol, which dissolves it readily without dissolving the main product.

The material thus prepared was obtained in the form of a light-green, granular powder.

having a constant melting-point of 234°C. (uncor.)
The substance contains no chlorine.

0.2014 gram gave 15.81 cc N.
0.1970 "   "   15.67 "   "
0.2075 "   "   0.1756 gram BaSO₄
0.2011 "   "   0.1692   "   "

Calculated for $C_6H_3NO_2 \langle{}^{CO \cdot C_6H_5}_{SO_2 \cdot N}\rangle{} n$ Found
                                                    1   2
N   9.72   9.86   9.99
S   11.11   11.60   11.55

The results of analysis, together with those described in the following section, indicate that the main product of the action of concentrated ammonia on the chloride is the lactim of paranitroorthobenzoylbenzenesulphonic acid. The reaction by which it is formed is represented as follows:

$C_6H_3NO_2\langle{}^{CO \cdot C_6H_5}_{SO_2 \cdot Cl}\rangle{} + 2NH_4OH = C_6H_3NO_2\langle{}^{C \cdot C_6H_5}_{SO_2 \cdot N}\rangle{}n + HCl + 2H_2O$

The lactim is insoluble in water, only slightly soluble in alcohol and readily soluble in benzene.

the formation of the lactim of the sulphonic acid by the action of concentrated ammonia agrees with the formation of the lactim of orthobenzoyl-benzenesulphonic acid by the action of dry ammonia gas on the sulphone chloride as observed by Saunders.[1]

XIII. The Action of Concentrated Ammonia on the Lactim of Paranitroorthobenzoylbenzene-sulphonic Acid.

The presence of a red colored amorphous product, melting above 275°C. (uncor.), with the lactim formed by the action of ammonia on the chloride, together with the fact that the amount of this product was increased as the length of time of heating was increased, indicated that another product was formed by the continued action of ammonia. A considerable quantity

[1] A. C. J. XVII 359.

of this material was prepared by heating some of the paranitroorthobenzoylbenzene sulphone chloride in a sealed tube until the only product consisted of the red-colored substance desired. It was found necessary to heat it to the temperature of the water-bath for twenty four hours in order to effect this transformation, while two and a half hours were sufficient to transform the sulphone chloride into the lactim.

The product is insoluble in water, but dissolves readily in absolute alcohol, giving a red solution with a marked green fluorescence. It is thrown out of solution by adding a considerable volume of water. On evaporating the alcoholic solution it is deposited as a red-colored crust which seems to possess no crystalline structure.

0.2150 gram gave 9.24 cc N = 1.24 per cent. N.

0.1615 gram gave 0.1200 gram $BaSO_4$ = 10.20 per cent. S

The results of the analyses show that, while the percentage of sulphur remains about the same as in the lactim, the percentage of nitrogen is increased, but not to an amount corresponding to the composition of any substance likely to be derived from the lactim by the further action of ammonia.

These results, together with the impossibility of obtaining the product in crystalline condition, and its properties generally, indicate that it is probably not a definite chemical compound, and that the lactim probably undergoes decomposition by the further action of ammonia.

XIV. The Action of Dilute Hydrochloric Acid on the Lactim of Paranitroorthobenzoyl benzenesulphonic Acid.

The action of hydrochloric acid on the lactim was first tried by boiling in a flask provided with a return-condenser, with an excess of acid. The lactim showed but little change after boiling with the acid for thirty hours. The acid was colored yellow, but this was found to be due to the solution of the lactim. By evaporating off the acid, the lactim is recovered with its melting-point unchanged. By heating the lactim with a large excess of hydrochloric acid in a closed tube to 150-160°C in a furnace for five hours, about half of the lactim is dissolved and is not deposited on cooling. On heating to 200°C. for seven hours longer, all of the lactim

goes into solution, and is not deposited on cooling, and the acid has a dark yellow color. On evaporating the filtered acid solution, a yellow, crystalline product is obtained which has not a constant melting-point. The melting-point, immediately after pressing out between filter-paper, is 00–160°C (uncor.) and it is charred by heating to 210°C (uncor.) in an air bath.

If, however, it is first carefully dried, it appears to melt at a much higher temperature. This indicates that the product is a salt which melts in its water of crystallization.

0.2009 gram gave 14.33 cc N.

0.1692 gram gave 0.1264 gram $BaSO_4$

Calculated for $C_6H_3NO_2\genfrac{}{}{0pt}{}{COC_6H_5}{SO_2ONH_4}$		Found	
N	8.64		8.95
S	9.88		10.24

The analysis is of a sample carefully dried.

The results of analyses indicate that the product is the ammonium salt of paranitro-orthobenzoylbenzenesulphonic acid. The transformation takes place according to the following equation:

$$C_6H_3NO_2 \genfrac{<}{}{0pt}{}{C\cdot C_6H_5}{SO_2\cdot n} + 2H_2O = C_6H_3NO_2 \genfrac{<}{}{0pt}{}{COC_6H_5}{SO_2\cdot ONH_4}$$

The ammonium salt thus obtained has generally the form of a yellow crystalline powder, but, under the conditions existing in one of the experiments, a few thick, needle-shaped crystals about a centimeter long were obtained. It dissolves readily in water.

XV. Conclusion.

The principal results obtained in the foregoing investigation may be briefly stated as follows:

By using phosphorus pentachloride in the proportion of two and a half molecules to one of the anhydrous acid potassium salt and heating for five hours under the conditions indicated, the unsymmetrical chloride is the only product. This may be crystallized readily in any quantity at the temperature of the laboratory by using ligroin (90-125°C.) as the solvent, provided the impurities are first removed by washing with ligroin (50-80°C.)

The action of benzene and aluminium chloride on the symmetrical and on the unsymmetrical chloride gives, in both cases, paranitroorthobenzoibenzenesulphone chloride.

The action of hydrochloric acid, concentrated

or dilute, dilute sulphuric acid, water and alcohol on paranitroorthobenzoylbenzenesulphone chloride is the same. The product formed is, in each case, paranitroorthobenzoylbenzenesulphonic acid which forms a series of characteristic salts.

The action of phosphorus pentachloride on the sodium salt of paranitroorthobenzoylbenzenesulphonic acid gives rise to the formation of paranitroorthobenzoylbenzenesulphone chloride identical with that from which the acid was derived by the action of acids or water.

The action of concentrated ammonia on paranitroorthobenzoylbenzenesulphone chloride for a limited length of time gives the lactam of paranitroorthobenzoylbenzenesulphonic acid.

The further action of concentrated ammonia gives a substance of indefinite composition which probably indicates a decomposition of

the lactim which is first formed.

The continued action of concentrated hydrochloric acid at a high temperature in a sealed tube converts the lactim into the ammonium salt of paranitroorthobenzoylbenzene sulphonic acid.

Biographical

The author was born at Dedham, Mass., August 26th 1867. His early education was received in the public schools of Newton, Mass. In 1885 he entered the Massachusetts Institute of Technology and received the degree of Bachelor of Science in 1890. From 1890 to 1893 he was engaged in experimental work in connection with the water supply of Boston, Mass. In 1893 he entered the Johns Hopkins University, where he has since pursued courses in chemistry, mineralogy and geology.

www.ingramcontent.com/pod-product-compliance
Lightning Source LLC
Chambersburg PA
CBHW030400170426
43202CB00010B/1443